Tucholsky Wagner Zola Scott Sydow Freud Schlegel
Turgenev Wallace Fonatne
Twain Walther von der Vogelweide Fouqué Friedrich II. von Preußen
Weber Freiligrath Frey
Fechner Fichte Weiße Rose von Fallersleben Kant Ernst Frommel
Richthofen
Hölderlin
Engels Fielding Eichendorff Tacitus Dumas
Fehrs Faber Flaubert Eliasberg Ebner Eschenbach
Feuerbach Maximilian I. von Habsburg Fock Eliot Zweig Vergil
Ewald
Goethe London
Mendelssohn Balzac Shakespeare Elisabeth von Österreich Dostojewski Ganghofer
Lichtenberg Rathenau Doyle Gjellerup
Trackl Stevenson Tolstoi Hambruch
Mommsen Thoma Lenz Hanrieder Droste-Hülshoff
von Arnim Hägele Hauff Humboldt
Dach Verne Rousseau Hagen Hauptmann
Karrillon Reuter Garschin Baudelaire Gautier
Damaschke Defoe Hebbel
Descartes Hegel Kussmaul Herder
Wolfram von Eschenbach Darwin Dickens Schopenhauer Rilke George
Bronner Melville Grimm Jerome Bebel Proust
Campe Horváth Aristoteles Barlach Federer
Bismarck Vigny Voltaire Herodot
Gengenbach Heine
Storm Casanova Tersteegen Gilm Grillparzer Georgy
Chamberlain Lessing Langbein Gryphius
Brentano Claudius Schiller Lafontaine Iffland Sokrates
Strachwitz Bellamy Schilling Kralik
Katharina II. von Rußland Gerstäcker Raabe Gibbon Tschechow
Löns Hesse Hoffmann Gogol Wilde Gleim Vulpius
Luther Heym Hofmannsthal Klee Hölty Morgenstern Goedicke
Roth Heyse Klopstock Kleist
Luxemburg Puschkin Homer Mörike Musil
La Roche Horaz
Machiavelli Kierkegaard Kraft Kraus
Navarra Aurel Musset Lamprecht Kind Kirchhoff Hugo Moltke
Nestroy Marie de France
Laotse Ipsen Liebknecht
Nietzsche Nansen Ringelnatz
Marx Lassalle Gorki Klett Leibniz
von Ossietzky May vom Stein Lawrence Irving
Petalozzi Platon Knigge
Pückler Michelangelo Kock Kafka
Sachs Poe Liebermann
Korolenko
de Sade Praetorius Mistral Zetkin

The publishing house tradition has created the series **TREDITION CLASSICS**. It contains classical literature works from over two thousand years. Most of these titles have been out of print and off the bookstore shelves for decades.

The book series is intended to preserve the cultural legacy and to promote the timeless works of classical literature. As a reader of a **TREDITION CLASSICS** book, the reader supports the mission to save many of the amazing works of world literature from oblivion.

The symbol of **TREDITION CLASSICS** is Johannes Gutenberg (1400 – 1468), the inventor of movable type printing.

With the series, tradition intends to make thousands of international literature classics available in printed format again – worldwide.

All books are available at book retailers worldwide in paperback and in hardcover. For more information please visit: www.tredition.com

tradition was established in 2006 by Sandra Latusseck and Soenke Schulz. Based in Hamburg, Germany, tradition offers publishing solutions to authors and publishing houses, combined with worldwide distribution of printed and digital book content. tradition is uniquely positioned to enable authors and publishing houses to create books on their own terms and without conventional manufacturing risks.

For more information please visit: www.tredition.com

The 1893 Duryea Automobile In the Museum of History and Technology

Donald H. Berkebile

Imprint

This book is part of the TREDITION CLASSICS series.

Author: Donald H. Berkebile
Cover design: toepferschumann, Berlin (Germany)

Publisher: tredition GmbH, Hamburg (Germany)
ISBN: 978-3-8495-0462-5

www.tredition.com
www.tredition.de

Copyright:
The content of this book is sourced from the public domain.

The intention of the TREDITION CLASSICS series is to make world literature in the public domain available in printed format. Literary enthusiasts and organizations worldwide have scanned and digitally edited the original texts. tredition has subsequently formatted and redesigned the content into a modern reading layout. Therefore, we cannot guarantee the exact reproduction of the original format of a particular historic edition. Please also note that no modifications have been made to the spelling, therefore it may differ from the orthography used today.

Figure 1.—Duryea automobile in the Museum of History and Technology, from an 1897 photograph. The gear-sprockets were already missing when this was taken, and the chain lies loosely on the pinion. Shown at the right, the Duryea vehicle following the recent restoration (Smithsonian photo 34183).

Don H. Berkebile

THE 1893 DURYEA AUTOMOBILE

In the Museum of History and Technology

During the last decade of the nineteenth century a number of American engineers and mechanics were working diligently to develop a practical self-propelled vehicle employing an internal-combustion engine as the motive force. Among these men were Charles and Frank Duryea, who began work on this type of vehicle about 1892. This carriage was operated on the streets of Springfield, Massachusetts, in 1893, where its trials were noted in the newspapers. Now preserved in the Museum of History and Technology, it is a prized exhibit in the collection of early automobiles.

It is the purpose of this paper to present some of the facts discovered during the restoration of the vehicle, to show the problems that faced its builders, and to describe their solutions. An attempt also has been made to correlate all this information with reports of the now almost legendary day-to-day experiences of the Duryeas, as published by the brothers in various booklets, and as related by Frank Duryea during two interviews, recorded on tape in 1956 and 1957, while he was visiting the Smithsonian.

The Author: *Don H. Berkebile is on the staff of the Museum of History and Technology, in the Smithsonian Institution's United States National Museum.*

Of the numerous American automotive pioneers, perhaps among the best known are Charles and Frank Duryea. Beginning their work of automobile building in Springfield, Massachusetts, and after much rebuilding, they constructed their first successful vehicle in 1892 and 1893. No sooner was this finished than Frank, working alone, began work on a second vehicle having a two-cylinder engine. With this automobile, sufficient capital was attracted in 1895 to form the Duryea Motor Wagon Company in which both brothers were among the stockholders and directors. A short time after the formation of the company this second automobile was entered by the company in the Chicago Times-Herald automobile race on Thanksgiving Day, November 28, 1895, where Frank Duryea won a victory over the other five contestants—two electric automobiles and three Benz machines imported from Germany.

In the year following this victory Frank, as engineer in charge of design and construction, completed the plans begun earlier for a more powerful automobile. During 1896 the company turned out thirteen identical automobiles, the first example of mass production in American [Pg 4] automotive history. [1] Even while these cars were under construction Frank was planning a lighter vehicle, one of which was completed in October of 1896. This machine was driven to another victory by Frank Duryea on November 14, 1896, when he competed once again with European-built cars in the Liberty-Day Run from London to Brighton. The decision to race and demonstrate their autos abroad was the result of the company's desire to interest foreign capital, yet Frank later felt they might better have used their time and money by concentrating on building cars and selling them to the local market. Subsequently, in the fall of 1898, Frank arranged for the sale of his and Charles' interest in the company, and thereafter the brothers pursued separate careers.

Figure 2.—Workmen in the Duryea factory in Springfield, Mass., working on some of the thirteen 1896 motor wagons. (Smithsonian photo 44062.)

Frank, in 1901, entered into a contract with the J. Stevens Arms and Tool Company, of Chicopee Falls, Massachusetts, which built automobiles under his supervision. This association led in 1904 to the formation of the Stevens-Duryea Company, of which Irving Page was president and Frank Duryea was vice president and chief engineer. This company produced during its 10-year existence a number of popular and well-known models, among them a light six known as the Model U, in 1907; a larger 4-cylinder called the Model X, in 1908; and a larger six, the Model Y, in 1909. In 1914 when Stevens withdrew from the company, Frank obtained control. The following year he sold the plants and machinery, liquidated the company, and, due to ill health, retired.

Charles, in the meantime, located in Reading, [Pg 5] Pennsylvania, where he built autos under the name of the Duryea Power Company. [2] Here, and later in Philadelphia under the name of the

Duryea Motor Corporation and other corporate names, he continued for a number of years to build automobiles, vacuum cleaners and other mechanical devices. Until the time of his death in 1938, he practiced as a consulting engineer.

Figure 3.—Admittance card of C. E. Duryea to the U.S. Patent Office, 1887.
(Gift of Rhea Duryea Johnson.)

Early Automotive Experience

Figure 4.—Charles E. Duryea, about 1894,
as drawn by George Giguere from a photograph.
(Smithsonian photo 48335-A.)

Born in 1861 near Canton, Illinois, Charles E. Duryea had learned the trade of a mechanic following his graduation from high school, and subsequently turned his interests to bicycle repair. He and his brother James Frank, eight years younger, eventually left Illinois and moved to Washington D.C., where they were employed in the bicycle shop of H. S. Owen, one of that city's leading bicycle dealers and importers. While in Washington, Charles became a regular reader of the Patent Office Gazette, [3] an act which undoubtedly influenced his later work with automobiles. A short time later, probably in 1889, Charles contracted with a firm in Rockaway, New Jersey, to construct bicycles for him, but their failure to make delivery as promised caused him to go to Chicopee, Massachusetts, where he contracted with the Ames Manufacturing Company to do his work. Moving there in 1890, he obtained for his brother a position as toolmaker with the Ames Company. Thus, Frank Duryea, as he was later known, also became located in Chicopee, a northern suburb of Springfield.

During the summer, 1891, Charles found the bicycle business left him some spare time, and the gasoline- [Pg 6] powered carriages he had read of earlier came constantly into his mind in these periods of idleness. [4] He and Frank studied several books on gasoline engines, among them one by an English writer (title and author now unknown); [5] this described the Otto 4-stroke cycle as now used. Some engineers, however, were concerned because this engine, on the completion of the exhaust stroke, had not entirely evacuated all of the products of combustion. The Atkinson engine, patented in 1887, was one of the attempts to solve this as well as several other problems, thus creating a more efficient cycle. This engine was designed so that the exhaust stroke carried the piston all the way to the head of the engine, while the compression stroke only moved the piston far enough to sufficiently compress the mixture. The unusual linkage necessary to create these unequal strokes in the Atkinson engine made it seem impractical for a carriage engine, where compactness was desired.

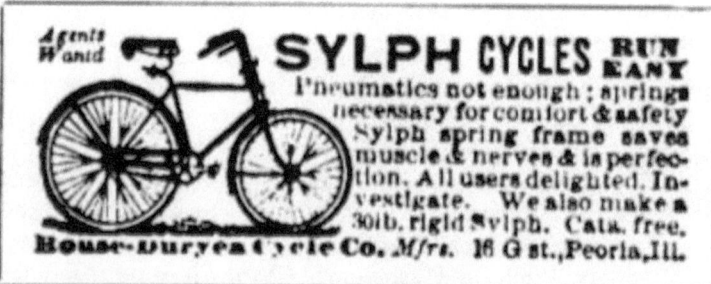

Figure 5.—Advertisement of Duryea bicycle company, *Scientific American*, September 9, 1893.

Going to Hartford, Connecticut, possibly on business relating to his bicycle work, Charles visited the Hartford Machine Screw Company where the Daimler-type engine was being produced, [6] but after examining it he felt it was too heavy and clumsy for his purpose. Also in Hartford he talked over the problem of a satisfactory engine with C. E. Hawley, an employee of the Pope Manufacturing Company, makers of the Columbia bicycle. Hawley, searching for a way to construct an engine that would perform in a manner similar to the Atkinson, yet would have the lightness and compactness necessary for a carriage engine, suggested an idea that Charles believed had some merit. This idea, involving the use of what the Duryeas later called a "free piston," was eventually to be incorporated in their first engine. [7]

Figure 6.—J. Frank Duryea, about 1894, as drawn by George Giguere from a photograph. (Smithsonian photo 48335.)

Construction Begins

Back in Chicopee again, Charles began planning his first horseless carriage. Frank later stated that they leaned heavily on the Benz patents in their work; [8] but while the later engine and transmission show evidence of this, only the Benz manner of placing the engine and the flywheel seem to have been employed in the original Duryea plan. Charles reversed the engine so that the flywheel was to the front, rather than to the rear as in the Benz patent, but made use of Benz' vertical crankshaft so that the flywheel rotated in a horizontal plane. Previously most engines had used vertical flywheels; Benz, believing that this practice would cause difficulty in steering a propelled [Pg 7] carriage, explained his reason for changing this feature in his U.S. patent 385087, issued June 26, 1888:

In motors hitherto used the fly-wheels have been attached to a horizontal shaft or axle, and have thus been made to revolve in a vertical plane, since the horizontal shaft is best adapted to the transmission of power. If, however, in this case we should use a heavy rotating mass, corresponding to the power employed and revolving rapidly in a vertical plane, the power to manage the vehicle or boat would become very much lessened, as the flywheel continues to revolve in its plane. I therefore so design the apparatus that its crank shaft x has a vertical position and its fly-wheel y revolves in a horizontal plane.... By this means the vehicle is not only easily controlled, but also the greatest safety is attained against capsizing.

To the Duryea plan, Benz may also have contributed the idea for positioning the countershaft, though its location is sufficiently obvious that Charles may have had no need for copying Benz. Charles wisely differed from Benz in placing the flywheel forward, thus eliminating the need for the long driving belt of the Benz carriage. Yet he did reject the bevel gears used by Benz, which might well have been retained, as Frank was later to prove by designing a workable transmission that incorporated such bevel gears. The initial plan, as conceived by Charles, also included the details of the axles, steering gear, countershaft with its friction-drum, the 2-piece angle-iron frame upon which the countershaft bearings were mounted, and the free piston engine with its ignition tube, since hot-tube ignition was to be employed. No provision was made, however, for a burner to heat the tube; nor had a carburetor been designed, though it had been decided not to use a surface tank carburetor. The plans called for no muffler or starting arrangement. [9] Many engines of the period were started simply by turning the flywheel with the hands, and Charles felt this method was sufficient for his carriage.

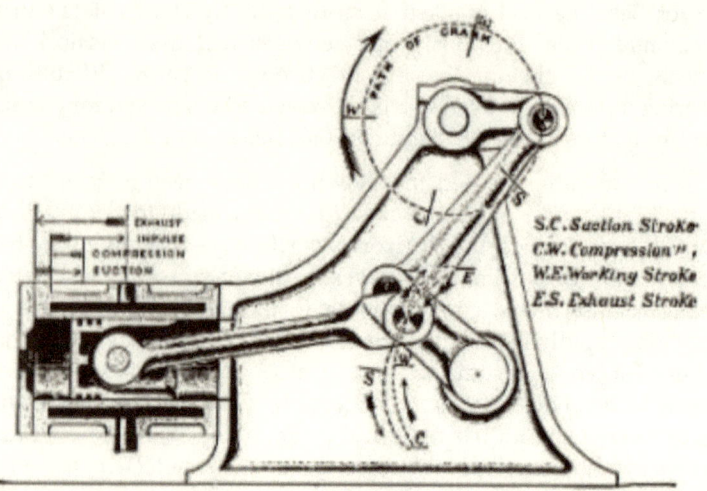

Figure 7.—Drawing showing principle of the Atkinson engine; this feature
is what the Duryeas were trying to achieve with their free-piston engine, by
substituting the free piston for the unusual linkage of the Atkinson.
(Smithsonian photo H3263-A.)

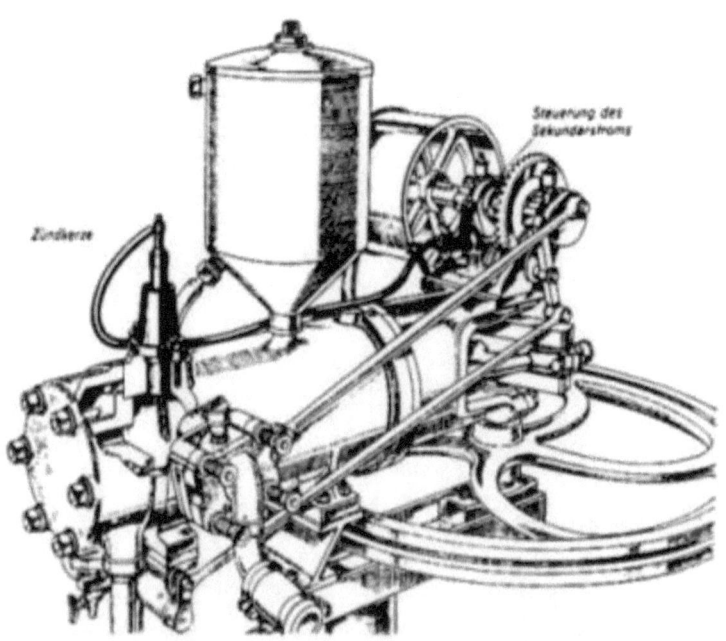

Figure 8.—Drawing of 1885 Benz engine, showing similarity in general appearance to Duryea engine. From Karl Benz und sein Lebenswerk, Stuttgart, 1953. (Daimler-Benz Company publication.)

The Ames plant customarily had a summer shutdown during August; thus, during August of 1891 Charles and Frank had access to a nearly empty plant in which they could carry on experiments and make up working drawings of the proposed vehicle. It cannot now be conclusively stated whether any parts were made for the car during August or the remainder of the year. It is more likely that the brothers attempted to complete a set of drawings. Frank Harrington, chief draftsman at Ames, may have helped out at this time; from Charles' statement of April 14, 1937, it is learned that he did prepare drawings during 1892.

[Pg 8]

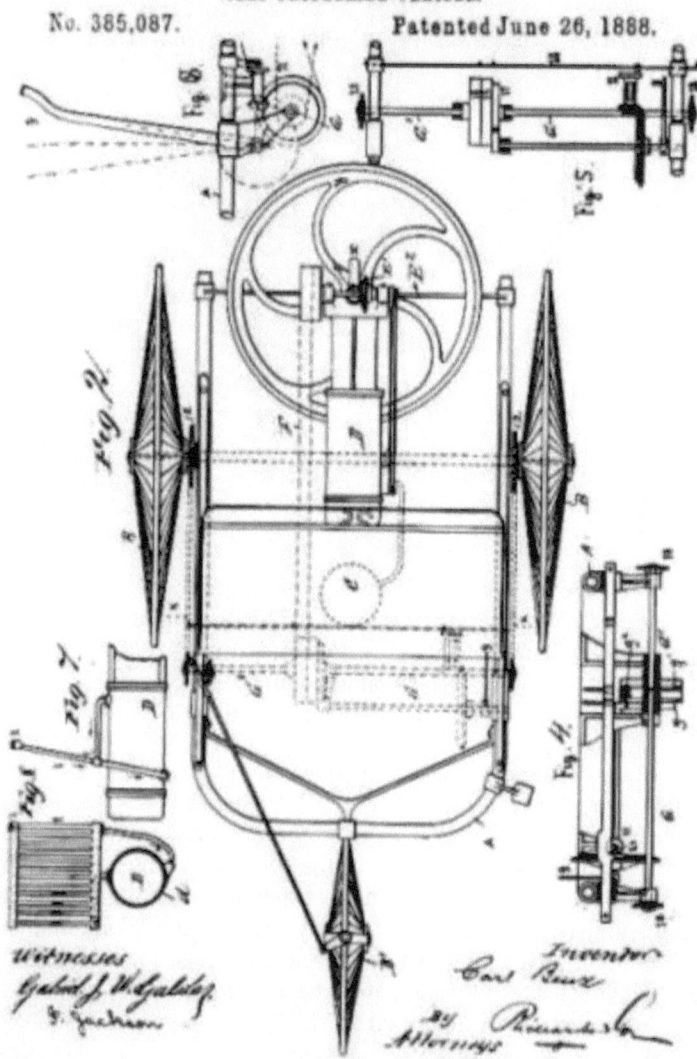

Figure 9.—Illustration from U.S. patent 385087, issued to Carl Benz, showing the horizontal plane of the flywheel, a feature utilized by the Duryeas in their machine.

The first contemporary record of any work on vehicles is a bill, dated January 21, 1892, for a drawing made by George W. Howard & Company. This drawing was made in the fall of 1891 by Charles A. Bartlett, a member of the Howard firm and a neighbor of Charles Duryea, according to a statement by Charles in the *Automobile Trade Journal* of Jan. 10, 1925. He was then also of the opinion that this drawing may not have had anything to do with the carriage they were about to assemble, but a notation found by Charles at a later date has led him to believe that it possibly concerned a business type vehicle he had discussed with an unidentified Mr. Snow.

By early 1892 Charles needed capital to finance his venture, an old carriage to attach his inventions to, a place to work, and a mechanic to do the work. On March 26, he stopped by the Smith Carriage Company and looked over a selection of used buggies and phaetons. He finally decided on a rather well-used ladies' phaeton which he purchased for $70. The leather dash was in so deplorable a state it would have to be recovered before the carriage went onto the road, and the leather fenders it once possessed had previously been removed; yet the upholstery appeared to be in satisfactory condition, and the candle lamps were intact.

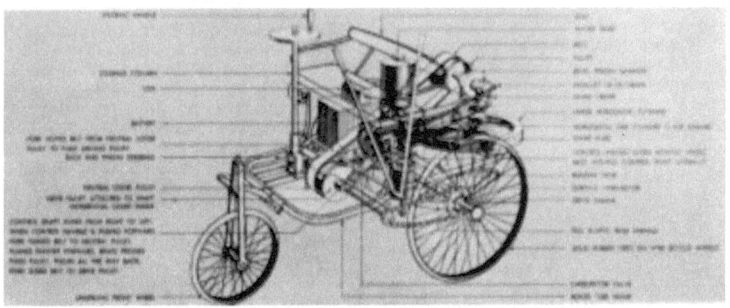

Figure 10. — Phantom illustration of Benz' first automobile.
(From *Carl Benz, Father of the Automobile Industry*, by L. M. Fanning, New York, 1955.)

Two days later, Charles was able to interest Erwin F. Markham, of Springfield, sufficiently to obtain his financial aid in the project. A contract was [Pg 9] drawn up between the two men, which stated that Mr. Markham was to put up $1000 for which he received a five-tenths share of the venture. When the $1000 had been used, he then had the option to continue his aid until the project had been carried to a successful climax, and retain his half share, or to refuse further funds and relinquish four of his five-tenths interest in the business. [10] Had he eventually chosen the latter, Charles would obviously have had to seek assistance elsewhere.

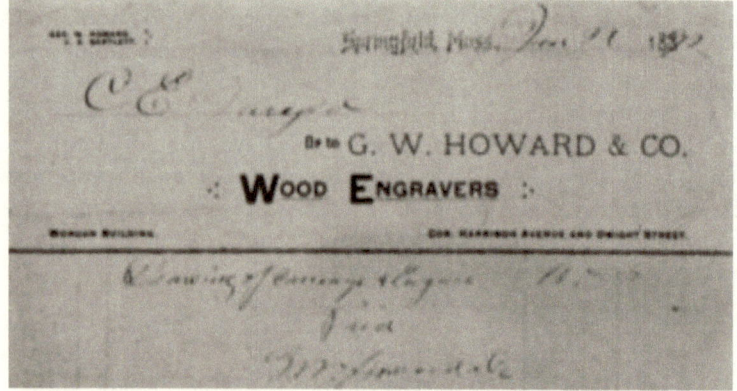

Figure 11.—The Howard & Co. bill showing the first work performed
toward a motor vehicle. While this may not refer specifically to the machine now in the museum, it is evidence of early work.

Figure 12.—The shop of John Russell & Sons.
It was on the second floor of this building that
Charles and Frank Duryea built their first motor vehicle.
(Courtesy of the *Springfield Union*.)

That same day, March 28, Charles found working space and machinery available at John W. Russell & Sons Company in Springfield. [11] The Russells had recently completed a large government order of shells for the famous dynamite guns later used on board the cruiser *Vesuvius* in the Spanish-American War, and this left an entire second floor, approximately 35 × 85 feet, virtually unoccupied, according to an affidavit of William J. Russell of April 30, 1926. Now ready to begin the actual work, Charles hired his brother Frank to start construction. Frank started about the first of April, receiving a raise of about 10 percent over the salary he had received at Ames. Before the vehicle was completed a number of other men performed work on some of the parts, among them William Deats who had been hired by Charles primarily to work on [Pg 10] bicycles in the same area, but who occasionally assisted on the carriage. Russell Company records show time charged against Charles Duryea by six other Russell employees: W. J. Russell, P. Colgan, C. E. Merrick, T. Shea, L. J. Parmelee, and A. A. Poissant.

Figure 13.—J. Frank Duryea looking over the Russell shop lathe on which he turned parts for the first Duryea vehicle. Photo taken about 1944. (Courtesy of the *Springfield Union*.)

It is Frank Duryea's remembrance that he started work on Monday, April 4. He first removed the body, with its springs, and placed it on a pair of wooden horses where it remained until the summer of the following year. The next step was to remove the rear axle and take it to a blacksmith shop where the old axle spindles were cut off and welded to a new drop-center axle. Following this the front axle spindles were removed, the ends of the axle slotted, and a webbed, C-shaped piece carrying the kingpin bearings was fitted into each slot, braced from underneath by short brackets which were riveted and brazed in place. The old spindles then were welded to the center of offset kingpins which in turn were mounted in their bearings in a manner similar to that in which the frame of the Columbia high-wheeled bicycle was mounted in its fork. Arms welded to the lower end of the kingpins were connected by the tie rods to an arm on the lower end of the vertical steering column, located on the center of the axle.

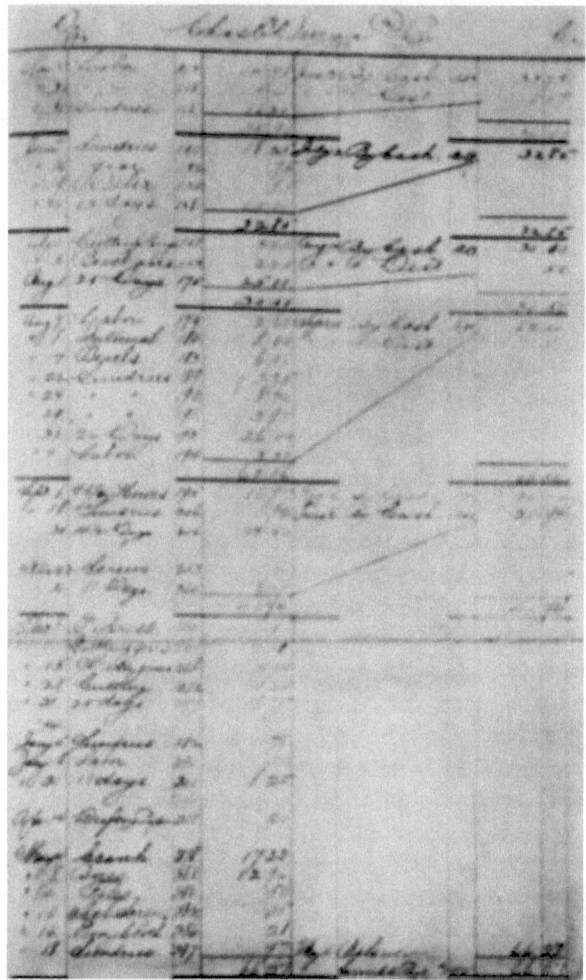

Figure 14.—A portion of the Russell shop records showing charges made against Charles Duryea during 1893-1894.

While work on the running gear advanced, some progress was made in the construction of the engine. [Pg 11] Patterns for the castings were fabricated, most of them by Charles Marshall on Taylor Street, [12] and castings were poured. The body or main casting of the engine resembled a length of cast-iron pipe: it had no bosses or lugs cast on, nor any water jacket, for they thought the engine would be kept cool merely by being placed in the open air. The front end of the engine was secured to the vehicle by four bolts which passed through the halves of the bearings and onto four projections on the open end of the engine. As the crankshaft of this engine was retained in constructing the present engine, it is logical to assume that the bearings were the same also. The head was cast as a thick disc, with both intake and exhaust valves located therein, and was bolted onto the flanged head end of the engine.

Inside the cylinder was the strange arrangement previously suggested by C. E. Hawley. To the connecting rod was attached a rather ordinary ringed piston, over which was fitted a free, ringless piston, machined to fit closely the cylinder bore. This floating piston could move freely a distance equal to the compression space. The intention was that on the intake stroke, suction would open the intake valve, which had no positive opening arrangement, and draw in the mixture which then was compressed as in a regular Otto engine. Fired by the hot-tube ignition system, the force of the explosion would drive both pistons down, forcing the outer one tight against the head of the smaller one, and at the end of the stroke the longer wall of the outer piston would strike an arm projecting into the cylinder near the open end, moving forward the exhaust valve rod to which the arm was attached, thus pushing open the valve in the head. [13] On the exhaust stroke the unrestrained outer piston moved all the way to the head, expelling all of the products of combustion and pushing the exhaust valve shut again. With a bore of four inches or less, this engine, Charles believed, should develop about three horsepower and run at a speed between 350 to 400 revolutions per minute. [14]

As no ignition system had yet been provided, they prepared a 4½-inch length of one-quarter inch iron pipe, closed at one end, and screwed the open end into the head. Heating this tube with an alcohol burner would cause ignition of the mixture when a portion of it was forced into the heated tube toward the end of the compression stroke. No attempt was made at this time to use the electrical make-and-break circuit used in their second engine, as the free piston would have wrecked the igniter parts on the exhaust stroke, and the push rod located on the end of the piston would have prevented the piston from closing the exhaust valve.

After keying the flywheel to the lower end of the crankshaft, Charles and Frank decided to make an [Pg 12] attempt to run the engine. Carrying it into a back room, probably during July or August, 1892, they blocked it up on horses. A carburetor had not yet been constructed, so they attempted to start the engine by spinning the flywheel by hand, at the same time spraying gasoline through the intake valve with a perfume atomizer previously purchased at a drugstore in the Massasoit House. Repeated efforts of the two men to start the engine resulted in failure.

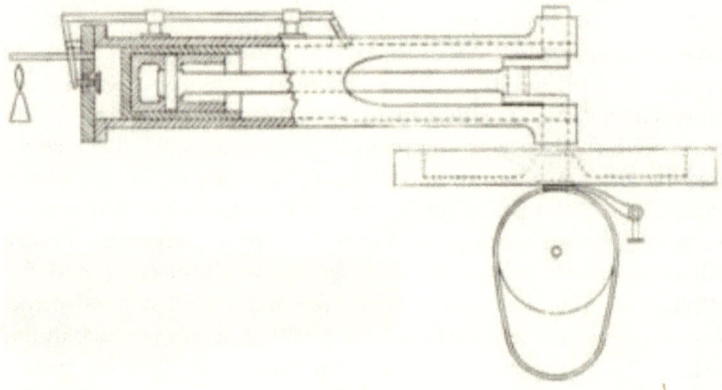

Figure 15.—Conjectural drawing of the free-piston engine used in the Museum vehicle prior to the present engine. (Drawing by A. A. Balunek.)

In the belief that the defects, whatever they might be, could be remedied after the engine was installed, the Duryeas went ahead and mounted the engine in the carriage. To do this they shortened the original reach of the carriage, allowing the engine itself to become the rear continuation of the reach. The four ears on the front, or open end of the engine, were bolted to the centrally located frame, with the bearing blocks in between. This frame, the same one now in the vehicle, was constructed of two pieces of angle iron, riveted and brazed together. Greater rigidity was obtained by a number of half-inch iron rods running from the frame to both front and rear axles. Because of the absence of any mounting brackets on the engine casting itself, a wooden block with a trough on top to receive the body of the engine was fitted between the engine and the axle, while two U-shaped rods secured it with clip bars and nuts underneath.

Beneath the flywheel was mounted the friction transmission of Charles' design. This consisted of a large drum, perhaps 12 inches in diameter, equal in length to the diameter of the flywheel and keyed to a shaft directly under the center of the crankshaft and parallel to the axles. (Diameter of drum estimated by examination of existing features.) In view of the four projections of the frame extending downward and just in front of the jackshaft position, it is likely that these supported the four jackshaft bearings. Being a bicycle manufacturer, Charles saw the need for a differential or balance gear. Accordingly, he purchased from the Pope Manufacturing Company a very light unit of the type formerly used on Columbia tricycles, and installed it somewhere on the jackshaft. A small sprocket on each end of the shaft carried a chain from the larger sprockets clamped to the spokes of each rear wheel. The lower surface of the flywheel had been machined so as to form a friction disc, with a one-quarter inch depression 3 inches in diameter turned in the center. The drum was positioned so that its upper surface was one-quarter inch below the face of the flywheel. Hanging loosely around the drum was an endless belt, one and one-half inches wide, first made of rather soft rubber packing material. The belt lay on the drum surface between the fingers of a shipper fork. While it lay under the 3-inch depression in the center of the flywheel, the belt and the drum were at rest, but when it was moved away from that depression the belt wedged itself tightly between the drum and flywheel, the resulting friction causing the drum to turn and setting the vehicle into motion. The farther the belt was moved toward the outer edge of the wheel, the faster the drum and the vehicle moved.

In September 1892, Charles, who had contracted with a Peoria, Illinois, firm to have bicycle parts manufactured, decided to move to that city. Departing on the 22d of September, he did not return to Springfield for over two years, and thus was not able to participate in the completion and testing of the [Pg 13] carriage. At the time of his departure several units on the carriage were incomplete. A carburetor had not been built, nor had a satisfactory burner or belt-shifting device. Charles had experimented with various shifting levers just before leaving Springfield: however, as he reported later, he did not succeed in designing a workable mechanism. [15] Frank Duryea, now left to finish the work unassisted, continued the experiments with the belt shifter. He finally worked out a fork mounted on a carriage that was supported by two rods, each of which slid in two bearings. Although the short distance between the two bearings caused the shifter carriage to bind occasionally, the device was thought to be sufficient and was installed just in front of the frame. Connected to a system of cables, arms, and rods, possibly similar to the present cam-bar shifter, the shipper-fork carriage was moved from side to side by raising or lowering the tiller.

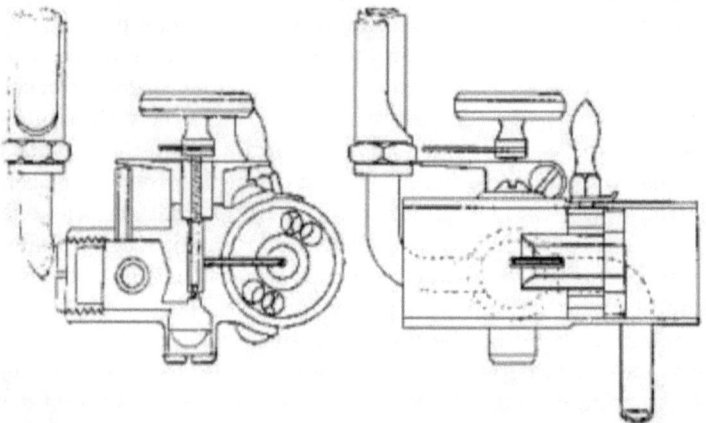

Figure 16.—Drawing of the carburetor used on both Duryea engines, 1893-1894,
showing sight feed on left and choke mechanism on right. (Smithsonian photo 13455.)

Turning now to an efficient burner for heating the ignition tube, Frank started with an ordinary wick-type kerosene lamp with a small metal tank. Wishing to use gasoline in the lamp, he found it necessary to fabricate a number of burner units before he found a type that gave him a clean blue flame. He then found the flame to be very sensitive to drafts and easily extinguished, and devised a small shield or chimney to afford it some protection.

Early in October, while still working with the burner, Frank developed a severe headache. He felt the fumes of the lamp had probably caused it, and went to his room in the home of a Mr. and Mrs. Patrick on Front Street in Chicopee. After he noticed no improvement, a doctor's examination showed he had typhoid fever, and on October 5 he was admitted to the Springfield Hospital. Here he remained for one month, being discharged on November 5. Returning to his room he was informed that because of the fear that he might be a typhoid carrier, the Patricks preferred him to find other lodgings. He readily accepted the invitation of Mr. and Mrs. D. H. Nesbitt of Chicopee to take a room with them. After several weeks recuperation in their home, he left Springfield to visit his mother in Wyoming, Illinois.

After a restful visit at home Frank Duryea returned to Springfield and finished the work on his burner. Now only the lack of a carburetor prevented a trial of the vehicle. He recalls that he studied several gasoline-engine catalogs and in one of them, a Fairbanks catalog he believes, [16] he saw a design that seemed to suit his needs. He decided to simplify the construction and operation of his carburetor and had a small bronze casting made to form the body of it. Inside was a gasoline chamber with two tapped openings, one to receive a pipe from the 2-gallon gasoline tank mounted above the engine, the other taking a pipe to the overflow tank underneath the engine, thus maintaining the gasoline level without the use of a float valve. This latter tank had a hand pump on one end so that the overflow gasoline could at times be pumped again into the main tank. Gasoline passed from the carburetor chamber through a needle valve, adjusted by a knob on top, then through a tiny tube that entered the pipe leading to the intake valve. It is not certain whether this intake pipe was at first fitted with the choke arrangement later used with the second engine.

Frank, hoping at last to be rewarded for his efforts by the sound of explosions from the engine, was ready to give the carriage an indoor trial. Standing astraddle of the reach and facing to the rear, he spun the flywheel with both hands, taking care not to get his hands caught between the wheel and the frame. His efforts were in vain, as there was complete failure to obtain ignition. He then made a new ignition tube, nearly twice as long as the original 4-½-inch tube, and turned down its wall as thin as he thought safety allowed. The thinner wall did not conduct the heat off so rapidly and thus kept the tube hot enough to permit ignition. After this slight change, he was able to get a few occasional explosions but he does not now believe that the engine ever operated continuously. Each explosion was accompanied by a [Pg 14] loud knock, due, undoubtedly, to the movement of the free piston. Had the engine operated continuously, it is likely that the action of the free piston would have shortly wrecked the engine. Further efforts appeared unwarranted until alterations could be made.

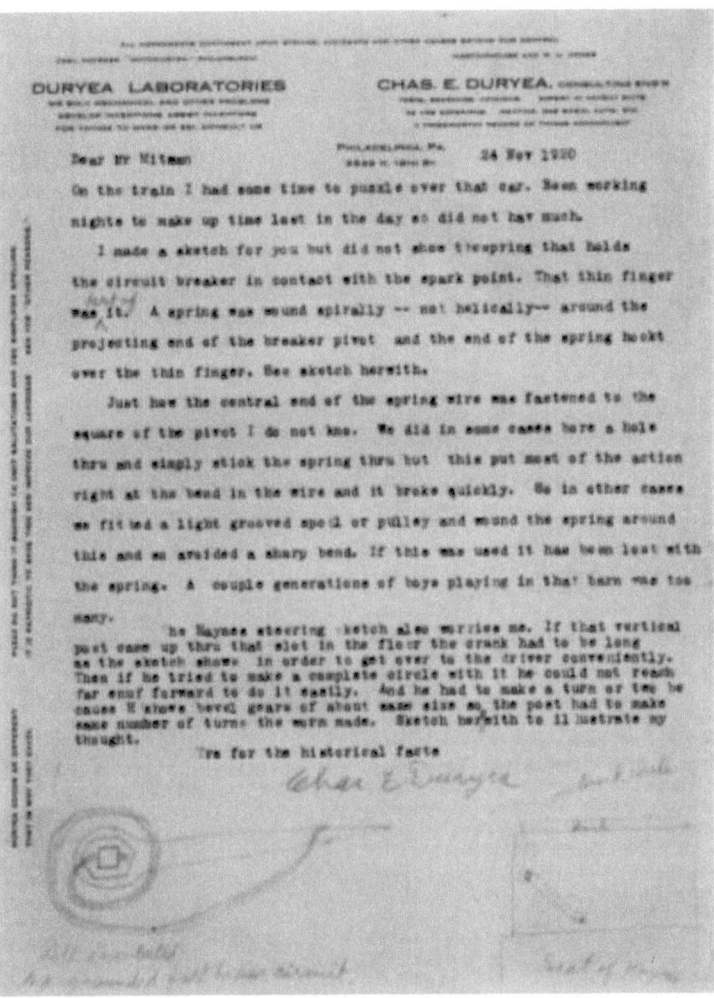

Figure 17.—Letter explaining the circuit breaker spring and the brass projection on top of the ignition chamber.
Mr. Mitman was, at the time, curator of engineering in the U.S. National Museum.

Text of Letter

[Pg 15] The two pistons were first pinned together into a single unit which was probably ringless, since it is believed the walls of the outer piston were too thin to admit rings. Because the piston no longer struck the exhaust valve, a short rod had to be screwed into the pistonhead; this pushed the valve shut at the completion of the exhaust stroke. The remaining problem, the opening of the exhaust valve, was solved by screwing a device to the side of the cylinder which operated from the sidewise motion of the connecting rod. This device shifted a small spacer between the piston and the striker arm of the exhaust-valve rod, permitting the piston to push open the exhaust valve. On alternating strokes the spacer shifted back out of the cylinder; therefore, no contact was made between piston and striker arm. Sometime in February 1893, the altered engine was successfully started.

At last the transmission could be tested. Will Russell had come upstairs to watch the trial, and according to a statement by him, given April 30, 1926, Frank, standing to the right of the engine and behind the rear axle, reached forward and with the combination tiller-belt-shifter, moved the belt into driving position. The carriage started forward, but as it approached the wall of the building Frank discovered that he could not get the belt back into the neutral position. In desperation, he grasped the rear axle with both hands and was dragged a short distance, attempting to stop the machine, before it struck the wall. He had, however, sufficiently retarded it so that no damage was done.

This short trial demonstrated some of the weaknesses in the friction transmission. Since the speed of the surface of the flywheel, in feet per second, increased in proportion to the distance of the point of contact from the center, the outer edge of the belt attempted to run faster than the inner edge. This conflict of forces not only put an undue load on the motor causing a great loss of power, but it also created a tendency for the belt to work towards the outer edge of the flywheel. Conversely, when the operator desired to return the belt to neutral, it strongly resisted any efforts to slide it toward the center of the wheel, as Frank had learned from the wall-bumping incident. Furthermore, the rubber belt on the friction drum had worn so badly that it had to be replaced at least once during the brief experiments.

Figure 18.—Ignition chamber, switch, and breaker contacts of the present Duryea engine.

At this point, Frank and Markham felt that the carriage was anything but satisfactory. While they were trying to decide what steps should be taken next, Frank added one last improvement to the engine. Fearing that the uncooled cylinder might suffer damage from the excessive heat, he constructed a copper water jacket in two halves, drew them together around the cylinder with clamping rings and soldered the seams. Asbestos packing sealed the end joints where the jacket contacted the cylinder. Thinking back, Frank does not recall that he ever used a water tank with this engine, though he does remember adding water through the upper jacket opening. The engine was run only for a few brief periods following this addition.

Obviously this collection of patchwork could not fulfill their needs for an engine. First, it would be next to impossible to start if the body was placed on the running gear, as the flywheel then would be practically inaccessible. The absence of rings on the piston caused a further loss of power to the already overloaded engine. The flywheel was too light. The absence of any form of governor left the operator with no control over the engine speed. Ignition was poor, partly owing to the hot-tube arrangement, and partly to the excessive distance between the engine and the carburetor. Frank wrote his brother Charles on February 6 [17] that in his opinion the mixing chamber was so far from the engine that the gasoline could not be drawn into the cylinder as liquid, and it was too cold to vaporize and go in as gas. Thus he had difficulty in getting the engine started. When it did start the explosions were unmuffled. Less important to him than these defects, however, was the awkward and unsightly wooden engine mount.

[Pg 16]

Description of the Automobile

Sometime in the early part of March, Frank convinced Markham that he could construct a new and practical engine, using only previously tried mechanical principles. [18] Drawing up new plans for this engine, he took them to Charles Marshall who began work on the patterns for the new engine castings. After the patterns had been delivered to the foundry, Frank left Springfield for a short vacation in Groton, Connecticut, where he visited with his fiancée. On May 17, 1893, several weeks after his return to Springfield, they were married.

The engine castings were undoubtedly received from the foundry prior to Frank Duryea's marriage, and the work of machining and assembling the parts went on through the spring and summer. This engine, still on the carriage in the Museum of History and Technology, is cased with a water jacket, and has bases on top to support the front and rear bearings of the starting crankshaft, and a base with port on the upper right side where the exhaust-valve housing was to be bolted. On the underside are two flanges, forming a base for seating the engine on the axle. A separate combustion chamber is cast and bolted to the head. Inside this chamber are located the igniter parts of Frank's electric ignition system. The fixed part, an insulated electrode, is screwed into the right side of the chamber and is connected with the ignition switch outside, to which one of the ignition wires is attached. A breaker arm inside is pinned to a small shaft extending through the top of the chamber. Around the breaker-arm shaft is a small coil spring (originally a spiral spring, according to the letter of Charles Duryea shown in fig. 17), anchored below to a thin brass finger extending toward the right side of the car, and above to a nut screwed tightly onto the shaft. This nut is also the terminal for the other ignition wire. The action of the spring keeps the breaker arm and the electrode in constant contact until the push rod on the end of the piston strikes the arm and separates the two parts. Breaking contact then produces the ignition spark. Since the mechanism would spark at the end of both the exhaust and compression strokes, the battery current is conserved by a contact strip, on the underside of the larger exhaust-valve gear, by means of which the flow of current is cut off during the greater part of the cycle.

On the left side of the combustion chamber is bolted the housing containing the tiny intake valve. A comparatively weak spring seats this valve in order that the suction created by the piston can easily pull it open. Clamped onto the valve housing is the intake pipe, enclosing the choke and carrying the carburetor on its forward side. The choke consists of two discs which block the pipe, each with four holes at the edges and one in the center. Turning one disc by means of a small handle outside, so that the four outer holes cannot coincide with those in the other disc, decreases the flow of air and causes all air to rush through the center hole, where the tiny carburetor tube passes through. The present carburetor was transferred over from the first engine. When Frank later installed the engine on the carriage he noticed the close proximity of the intake pipe to the open end of the muffler. Believing that the fumes might choke the engine, he attached a long sheet-metal tube to the intake pipe so that fresh air would be drawn in from a point farther forward on the vehicle.

Moving to the right side of the engine brings the exhaust-valve assembly into view. This valve is contained in a casting bolted over the exhaust port in the side of the cylinder, and from the casting a pipe leads to the muffler underneath. The valve is pushed open by a rod connected to a crank which is pinned to the lower end of a shaft carrying an iron gear on top. This gear is in mesh with a fiber gear, keyed to the upper end of the crankshaft, with half the number of teeth. This ratio permits the opening of the exhaust valve on every other revolution.

The crankshaft of the first engine was retained for the new engine, thus giving the two engines the same stroke of 5-3/8 inches, but the bore was increased slightly to 4 3/8 inches. With this larger bore and with the engine speed increased to 500 rpm, Frank rated this engine at 4 hp. [19] A heavier flywheel, with a governor resting in the upper recess, was pressed onto the crankshaft. As the operator of the vehicle had no control over the carburetor once he climbed into the seat, this governor was necessary to maintain regular engine speed. Its function was to move a slide on the exhaust-valve unit to prevent the valve from closing. Thus the engine, with the suction broken, could not draw a charge on the next revolution. During the recent restoration of this carriage it was found that while [Pg 17] most parts are still intact, nearly all of the governor parts are missing. A description of them must therefore be based on the recollections of Frank Duryea, along with certain evidences seen on the engine.

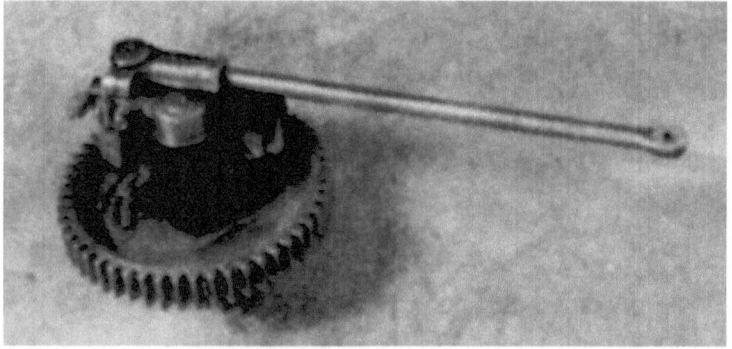

Figure 19. — Under side of exhaust valve mechanism showing electrical contacts that give spark only on every other revolution.

Just on top of the flywheel, and surrounding the crankshaft, rest two rings, 3 7/8 inches in diameter. Into the opposing surfaces of these rings are cut a series of small inclined planes, appertinent to each other. On the outer circumference of the upper ring two pins pass through a pair of lugs mounted in the flywheel, causing the ring to rotate with the flywheel, yet permitting vertical movement. Underneath, the other ring is allowed to turn slightly when, by means of two connecting links, the arms of the governor push against them. These two arms, each constructed like a right angle and pivoted at the apex, are arranged directly opposite each other far out in the flywheel recess. As a weight on one angle of the arm presses outward by centrifugal force against a spring, the other angle presses inward against the connecting link mentioned above. The turning of the lower set of inclined planes against the fixed set above raises the upper ring and the fork resting on it. The upward movement of this fork, which is a continuation of an arm pivoted to a bracket midway between the crankshaft and the slide carrying the exhaust valve stop, causes the other end of the arm to drop, pulling the slide down with it. In this manner the closing of the exhaust valve is blocked, preventing the intake of the next charge, and therefore the engine misses one or more explosions until it slows to its normal speed.

A starting shaft is mounted above the engine casting by a cast-iron bracket on either end. The front end of the shaft has a bevel gear which is held by a coil spring behind the front bracket, just out of contact with a bevel gear pressed onto the upper end of the crankshaft. The short rear portion of the shaft is a tube which slides over the main shaft. Fitting the removable handcrank to the squared end of the hollow shaft and turning the crank clockwise, will advance the forward section of shaft through the medium of a pair of inclined collars. With the bevel gears now engaged the engine may be cranked. When ignition begins, the inclined collars slide back down each other's surfaces, the shaft is again shortened, and its bevel gear springs free of the one on the crankshaft.

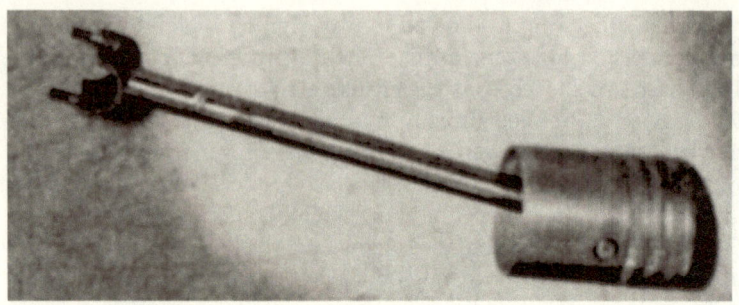

Figure 20. — Piston and connecting rod of second engine. Screw on rod is where oil is poured into connecting rod to lubricate wrist pin and crankshaft.

While Frank worked on his engine, he realized that certain parts of the old running gear would need to be altered or replaced. In view of the heavier and more powerful engine, he felt the old wheels, probably having compressed band hubs, were inadequate. He procured a set of new, heavier wheels [20] with Warner-type, cast-iron reinforced hubs. The angle iron frame, apparently sturdy enough to carry the added weight, was retained, but it was decided to install a heavier rear axle. [21] The front axle assembly was at first allowed to remain unchanged, as was the steering apparatus. A short time later when the engine and friction transmission were bolted in place on the running gear, Frank saw that the rigidity of the framework had an undesirable effect. When the vehicle passed over any unevenness in the shop floor, the framework was distorted and caused the jackshaft bearings to bind tightly enough on the shaft to prevent its being turned by hand. In order to provide the 3-point suspension necessary to eliminate this distortion, Frank attached the forward parts of the framework to an extra wooden spring bar, installing between this bar and the front axle a vertical fifth wheel of the type ordinarily used in a horizontal position in any light carriage.

[Pg 18] Frank next calculated that with the faster running engine the speed of the vehicle would be about 15 miles an hour, too much for the heavily loaded wheels. As he intended to make use of the original transmission, he decided to decrease the speed by increasing the size of the friction drum. He accomplished this by sliding a heavy fiber tube over the original drum, bringing its diameter to approximately 14 inches. The original shipper fork carriage was improved by separating the original bearings to a greater distance, and eliminating one of the two bearings on one end. This permitted a smooth and free operation of the small sliding carriage.

In August 1893, possibly as a result of indoor experiments, Frank discovered that the chains running from the small 5-tooth [22] jackshaft sprockets to the large, bronze, wheel sprockets were tight at some times and loose at others. This caused considerable unnecessary noise. The difficulty apparently was the result of the sprockets being cast and not machined. The patternmaker had said he believed he could make the pattern accurately enough so that no machining of the castings would be necessary. Nice castings were produced, but "these sprockets were the reason why an unusual construction was put on the crankshaft [meaning jackshaft]," explained Frank Duryea during an interview at the National Museum on November 9, 1956. Elaborating further, in reply to the queries of E. A. Battison, of the Museum's division of engineering, Duryea told of the problem and the solution when he explained that the sprockets had places where the shrinkage was not even. The hot metal, contracting as it cooled, did not seem to contract uniformly, creating slightly unequal distances between teeth. This resulted in the chain hanging quite loose in some places and in others the tightness prevented adjustment. He contacted Will Russell, foreman of the Russell shop, where the automobile was made, and Russell showed him a device, built by George Warwick, who had made the Warwick bicycle. It was an internal-cut gear, according to Duryea's description, with sprocket teeth on its periphery. With sprockets outside and normal teeth inside, the wheels were about 6 inches in diameter, externally.

These little internal-gear sprockets were hung on double-shrouded pinions secured to each end of the jackshaft. A solid disc or housing fitted against both ends of the pinion to prevent the internal gear from working off sideways. Duryea explained the function of these unique little parts: "as soon as tension came on that ring gear that we talked about, it not only tightened the chain hanging on this sprocket on the upper side, but it tightened it on both sides. [The sprocket] rocks right out: both sides of the chain are tight."

This feature is one rarely encountered elsewhere, and Duryea, later in the interview said, "To tell you the truth, I think I was just a little bit ashamed about the thing, because I had to pull it off. I didn't like the looks of it after I got it on."

Two small tanks, each with a capacity of approximately two gallons, were mounted over the engine in the positions they still occupy, the one on the left for gasoline, [23] the other for water. The small fitting under the gasoline tank has a thumbscrew shutoff and a glass-sight feed tube, leading to the carburetor. The water tank, an inch longer than the gasoline tank, communicates with the water jacket of the engine through two pieces of half-inch pipe, entering the jacket from above and below. The overflow tank, holding just over a gallon, is suspended between the rear axle and the flywheel.

A number of mufflers were constructed for the engine. [24] The first experimental one was built of wood, being a box 6 × 6 × 15 inches with a hole for the exhaust pipe in one end and a series of small holes in the opposite end. Inside, Frank arranged metal plates which were somewhat shorter than the depth of the box. Every other one was attached to the bottom of the box; the intermediate plates were fastened to the top. This contrivance muffled the sound considerably, but, as might be expected, soon began to smoke. There can be little doubt that it was replaced before any of the outdoor trials began. Another type consisted of a cylindrical metal shell, perhaps six inches in diameter and ten or twelve inches long. Here a series of perforated baffle plates were inserted, with alternating solid plates having parts of their external edges cut away. Two bolts running the length of the muffler held on the cast-iron heads in a manner quite similar to the Model-T Ford mufflers of later years. Though partially satisfactory, Frank, in a November 6, 1957, interview, complained that it [Pg 19] made a metallic sound. Perhaps this was the muffler he used from September to November 1893.

Figure 21.—Illustration of the no. 2 Samson battery used by the Duryeas in their vehicle. (Smithsonian photo 46858.)

On August 28 Frank wrote to Charles saying the carriage was almost ready for the road and that he hoped to take it out for a test on the coming Saturday, "off somewhere so no one will see us...." [25] There is no evidence showing whether the amount of remaining work permitted the proposed trial on September 2. The body was finally replaced on the running gear, at which time it was found necessary to raise the seat cushion several inches by the insertion of a framework made of old crating boards. This allowed sufficient room between the seat and the frame to suspend the batteries and coil. Six no. 2 Samson batteries were contained in this space, three on each side, in rows parallel to the side of the vehicle. The Samson battery consisted of a glass jar containing a solution of ammonia salts and water, with a carbon rod in the center, housing a zinc rod. It is difficult to understand why they used Samson batteries rather than dry cells; perhaps they were concerned with the mounting cost of the machine and were making use of parts already on hand. [26] A coil, possibly from an old gaslight igniter system, accompanied the Samson batteries under the seat. This original coil is now missing.

The iron dash frame, previously recovered and provided with a rain apron to be pulled up over the knees in the event a heavy rain blew in under the carriage top, was bolted back in place. Frank and Mr. Markham gave the carriage a quick painting; later Frank admitted, "the machine never had a good job of painting." [27] Before the motor wagon actually got onto the road, a reporter on the *Springfield Evening Union* got some statistics on it and an item appeared on September 16, giving the first public notice of the machine.

[Pg 20]

NO USE FOR HORSES.

Springfield Mechanics Devise a New Mode of Travel.

Ingenious Wagon Now Being Made in This City for Which the Makers Claim Great Things.

A new motor carriage, which, if the preliminary tests prove successful as is expected, will revolutionize the mode of travel on highways, and do away with the horse as a means of transportation, is being made in this city. It is quite probable that within a short time one may be able to see an ordinary carriage in almost every respect, running along the streets or climbing country hills without visible means of propulsion. The carriage is being built by J. F. Duryea, the designer and M. F. Markham, who have been at work on it for over a year. The vehicle was designed by C. E. Duryea, a bicycle manufacturer of Peoria, Ill., and he communicated his scheme to his brother, who is a practical mechanic in this city.

The propelling power is furnished by a two-horse power gasoline motor situated near the rear axle and which, when started, runs continuously to the end of the trip, notwithstanding the number of times the carriage may be stopped. The speed of the motor is uniform, being about 500 revolutions a minute, and is so arranged that it gives a multiplied power for climbing hills and the lower the rate of speed the greater power is furnished by the motor. The slowest that the carriage can be driven is three miles an hour and the speed can be increased to fourteen or fifteen miles an hour. The power is transferred from the driving wheel of the motor, which runs horizontally with the main shaft by an endless friction belt running on a drum wheel. The belt is controlled by a lever within easy reach of the driver and is shifted along the drum wheel to increase or decrease the speed. The driving wheel is about twenty inches in diameter, having in its center a depression to which the belt is shifted to stop the carriage.

The carriage can be reversed by shifting the belt from the end of the drum, which gives the forward motion to the opposite side beyond the depression in the driving wheel. The power which has been transferred to the driving shaft from the motor is in turn transferred to the two rear wheels of the carriage by a combination gear and sprockets. An endless chain connects the sprockets on the carriage wheels to the sprocket wheels on the driving shaft. All of the motive power is located under the body of an ordinary phaeton, the hight of which is not increased by the machinery. The motor is started by a crank which is easily applied to a shaft in the rear of the carriage and the gasoline is ignited in the cylinder by electricity. An automatic device stops the flow of gasoline into the cylinder when the motor ceases running. The gasoline is carried in tanks, which hold about two gallons, and which will run the carriage for about eight hours. The wagon is guided by a bicycle bar, and the speed is also controlled by this bar.

The method employed in this is as follows: To start the carriage press the lever down; to reverse it throw the lever up and to guide the wagon turn the lever either to the right or left. The front axle instead of turning horizontally plays up and down, in order that the machinery may be on a level with the rear wheels, while the front wheels are set on the axle by a pivotal joint and are connected with the guiding lever by bars with ball bearings. The carriage complete weighs about 250 pounds, and the essential features are already covered by patents while others are pending.

It is estimated that the carriages can be sold for about $400, and a stock company will probably be formed to manufacture them.

Figure 22. — From the *Springfield Evening Union*, September 16, 1893. Text of Article

Toward the latter part of the following week, Frank was ready to give the product of his labors its first road trial. On September 21 the completed carriage was rolled onto the elevator at Russell's shop. Seeing that the running gear was too long for the elevator, they raised the front of the machine, resting the entire weight of 750 pounds on the rear wheels. Once outside the building, they pushed it into an area between the Russell and Stacy buildings. After dark, "so no one will see," Will Bemis, Mr. Markham's son-in-law, brought a horse and they pulled the phaeton out to his barn on Spruce Street. [28] There, on Spruce and Florence Streets the first tests were made. The next day Frank wrote his brother saying, "Have tried it (the carriage) finally and thoroughly and quit trying until some changes are made. Belt transmission very bad. [29] Engine all right." He did admit the engine seemed to be well loaded most of the time. He also had an idea in mind to replace the poor trans [Pg 21] mission, explaining the plan to Charles: "The three gears [30] on secondary shaft have friction clutches, the two bevel gears on same shaft are controlled by a clutch which frees one and clutches the other at will. This provides a reverse."

Figure 23. — Type of spark coil the Duryeas are believed to have used in their electrical circuit,
as shown in a catalog illustration. (Smithsonian photo 46858-A.)

The *Springfield Evening Union* of September 22 carried a notice of the trial. This report, too, commented on the faulty transmission and the plan already in Frank's mind for the new transmission.

... The friction belt allowed of the speed being steadily increased or diminished at the will of the driver and caused no sudden forward motion of the carriage, but while this arrangement has many advantages it uses up the power so that the two-horse power furnished by the motor [somewhat less than the rating Frank gave the engine] was reduced to less than three-fourths horse power on reaching the main shaft. This would not be sufficient to propel the carriage up steep grades but would be sufficient to run the carriage on level road.

The inventors will do away with this belt in favor of a clamp gear and will make the drum wheel smaller. By this means there will be very little power lost in transmission to the shaft and by a patented arrangement the carriage may be started gradually but the speed must be increased by shifting the clamp gear to a succession of gears on the driving wheel of the motor. The speed of the carriage will be fixed permanently according to the size of the gear that the smaller one is shifted to. The test of the machine with the gear arrangement will be made soon.

In October Frank decided on another vacation and went to Chicago to see the Columbian Exposition. Charles had come up from Peoria to see the fair and the two talked over the progress on their motor wagon, and discussed the transmission problem. They gave particular attention to everything relating to engines and motor carriages, and Frank recalls seeing a Daimler quadricycle that impressed him with its performance. [31] Just what decisions the two might have made there are unknown, yet it is likely that they agreed to give the old transmission one more chance to prove itself.

Returning to Springfield, probably in the first week of November, Frank gave the friction drive its final test, this time substituting a leather belt for the rubber one first used. [32] Mr. Markham, though intensely interested in the experiments, apparently was dubious concerning the safety of the carriage. It had no brakes, and fearing failure of the transmission on a downgrade, he was reluctant to ride in the machine. On November 9 he asked Will Bemis to try it for him. The following day the *Springfield Morning Union* gave a description of the run:

Residents in the vicinity of Florence street flocked to the windows yesterday afternoon astonished to see gliding by in the roadway a common top carriage with no shafts [Pg 22] and no horse attached.

The vehicle is operated by gasoline and is the invention of Erwin Markham and J. F. Duryea. It has been previously described in The Union and the trial yesterday was simply to ascertain the practical value of a leather friction surface which has been substituted for the rubber one previously used. The vehicle, which was operated by Mr. Bemis, started from the corner of Hancock avenue and Spruce street and went up the avenue, up Hancock street and started down Florence street, working finely, but when about half-way down the latter street it stopped short, refusing to move. Investigation showed that the bearing had been worn smooth by the friction and a little water sprinkled upon it put it in running condition again. The rest of the trip was made down Florence and down Spruce street, to the residence of the inventors. They hope to have the vehicle in good working condition soon.

Figure 24.—Running gear of Duryea vehicle, showing the second engine and other parts as used in January 1894.

The same evening, the late edition ran a brief paragraph stating that "the test was made to determine the value of a leather friction surface for propelling the wagon, that had been substituted in place of the rubber surface, used in the former test." Bemis, according to

Frank Duryea's recollection, was not impressed with the performance of the machine, saying "the thing is absolutely useless," and for a time it appeared that further support from Markham would not be forthcoming. Frank, believing eventual success to be near, drew up plans showing his geared transmission, and with these managed to gain Markham's partial support. Money for material and use of the shop was to continue, but Frank was to complete the work on his own time.

Now receiving no salary, Frank worked hurriedly on the transmission throughout late November, December, and the first two weeks of January. First discarding the old friction drum and shaft, and the shipper-fork carriage, he bolted a rawhide bevel gear to the lower surface of the flywheel. This turns two bevel gears, in opposite directions, on a countershaft directly underneath, approximately in the position of the old jackshaft. The right bevel gear is secured to the main countershaft on which two clutches are mounted, one on each side of the crankshaft. On a sleeve turning freely around the countershaft is mounted the reverse bevel gear and clutch. Three free-running clutch drums, the right one carrying the high-speed gear, the two on the left carrying the combination low speed and reverse gear between them, complete the counter [Pg 23] shaft assembly. The clutch assemblies are of Frank Duryea's design, having internal arms, expanding outward to press leather-faced shoes against the inner surface of the drum, thus securing the drum and its gear to the shaft. Behind this machinery is the jackshaft with its small differential on the right, two laminated rawhide gears [33] meshing with the iron gears of the countershaft, and the internal-gear sprockets hanging on the small pinions at either end. A sliding cam bar, mounted nearly in the position of the former shipper-fork carriage, is operated by the vertical movement of the tiller handle to engage any one of the three clutches. With the tiller depressed, the vehicle is in reverse. Elevating it slightly puts it into low gear, and raising it still higher runs the machine at its highest speed.

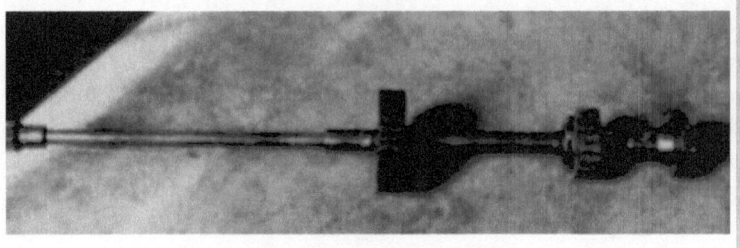

Figure 25.—Half of Jackshaft, showing rawhide gears, double shrouded pinion and half of the Columbia differential.

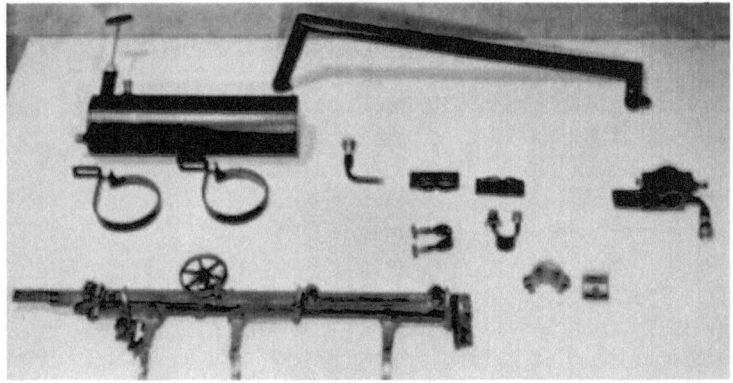

Figure 27.—Cam bar in foreground, operated by tiller, actuates the various clutches of the transmission. The overflow gasoline tank with the hand pump can be seen in the rear.

[Pg 24]

United States Patent Office.

CHARLES E. DURYEA, OF PEORIA, ILLINOIS.

ROAD-VEHICLE.

SPECIFICATION forming part of Letters Patent No. 540,648, dated June 11, 1895.

Application filed April 20, 1894. Serial No. 509,485. (No model.)

To all whom it may concern:

Be it known that I, CHARLES E. DURYEA, a citizen of the United States, residing at Peoria, in the county of Peoria and State of Illinois, have invented new and useful Improvements in Road-Vehicles, of which the following is a specification.

The object of this invention is to produce a road vehicle which shall be self-propelled, not unduly heavy, simple and easy of control and comparatively inexpensive, together with such minor objects as will become hereinafter apparent.

The invention more particularly relates to the construction and arrangement of parts for constituting the driving gearing and to the means for controlling the action thereof; to an improved manner of mounting the front, or steering, wheels upon the front axle, and of mounting the said axle relative to the running gear frame, and to the means for effecting the steering; to the appliances for the support of the motor and driving mechanism in an advantageous and efficient manner, and, generally, to improved and simplified details of construction throughout the vehicle, all as will hereinafter be rendered more apparent, and the invention consists in constructions and combinations of parts, all substantially as will hereinafter fully appear and be set forth in the claims.

Reference is to be had to the accompanying drawings, in which—

Figure 1 is a sectional elevation from front to rear of the improved road-vehicle. Fig. 2 is a plan view of the running and driving gear, the vehicle-body being understood as removed. Fig. 3 is a front elevation of the vehicle. Fig. 4 is a perspective view of the support and suspension devices for the driving mechanism. Fig. 5 is a vertical sectional view, longitudinally, through the shiftable driving-gear, the controlling devices employed in conjunction with this mechanism being seen in side elevation. Figs. 6 and 7 show the above-mentioned controlling devices in operative relations differing the one from the other and also from that of Fig. 5.

Similar letters of reference indicate corresponding parts in all of the views.

The parts will now be described in detail with reference to said drawings, and A represents the body which is spring supported on the frame, B, of the running gear. This frame, as shown, is rectangular, and has the body-supporting springs, B², similar to those found in common carriages. This frame has, affixed thereto, at its rear ends, sleeves, a, a, which loosely embrace the rear wheel axle, D, which is the driven axle of the vehicle. The axle, E, for the front wheels is centrally secured to the running gear frame, B, by the horizontal king-bolt, b, whereby such axle may have a swinging movement relative to the frame in a vertical plane, but it has no swinging movement horizontally, the wheels being swivel-mounted on the ends of this axle peculiarly, as will shortly hereinafter be set forth.

The body, as shown, is in the form of an inverted box, the motor, H, and driving gear being accommodated within the downwardly opening inclosure constituted thereby, and the body also has the upwardly open box-like forward extension, or pit, A², for the accommodation of the feet of the rider, the rider's seat being constituted by the top forward portion of the box body. Some other suitable design of body may, of course, be used in lieu of this one shown.

The front wheels, d, d, are hung to the front axle, E, so that the center of each wheel base is in a line coincident with the axis of the pivotal connection which is provided between the journals for the wheels and the axle, which arrangement practically destroys any tendency to deflection from the course that might otherwise arise from striking an obstacle, and so renders the steering easier. In order to effect this the axle is formed with yoked ends, the yoke members, f, f, being above and below the longitudinal line of the axle. The short journal, g, shown for each wheel, has at its inner end an upwardly and downwardly extended arm, h, which is return-bent to be loosely embraced by the axle yoke, f, f. The cone pointed screws, c, passed through the yoke members, f, and into sockets therefor in the arms, h, of the journals, g, constitute the means for the swivel connection between said parts. The lock-nuts, c², manifestly, are employed with utility in this connection.

It will be perceived that inasmuch as in the arrangement shown, the pivotal connections

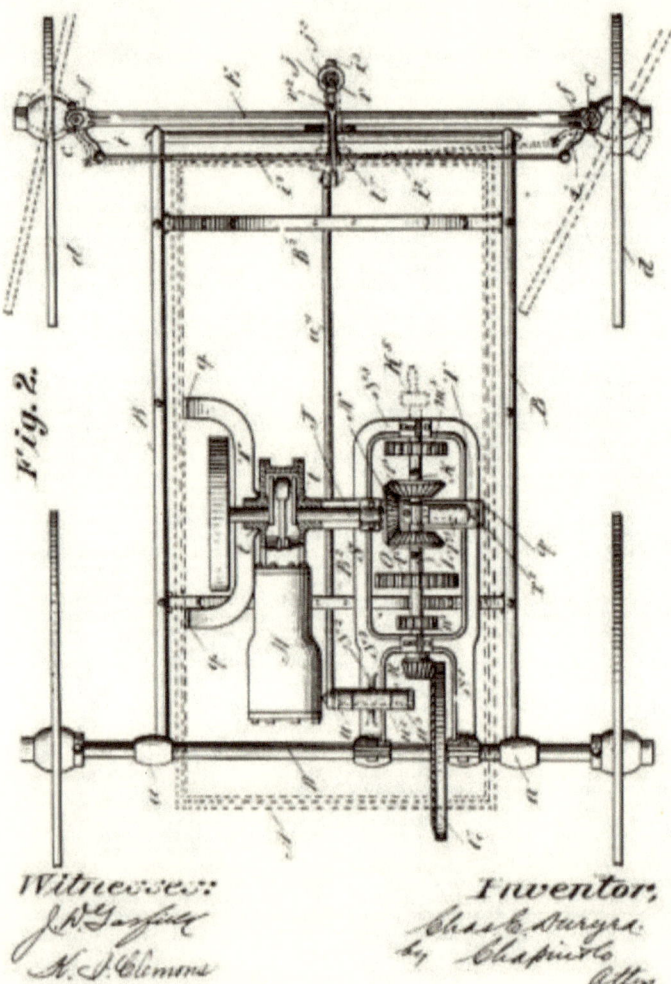

Figure 28.—A drawing and the first page of the specifications of the first patent issued to C. E. Duryea.

It can be readily seen that this drawing was not made after the plan of the first vehicle.

Text of Patent Letter

[Pg 26] As the work moved nearer completion Frank realized that the final tests would have to be conducted on roads made icy by falling snows. He had considerable doubt whether the narrow iron tires would have enough traction to move the phaeton. Soon he devised an expedient for this situation, communicating to Charles on December 22 that he was "having Jack Swaine [a local blacksmith] make a couple of clutch rims so we can get over this snow and ice.... Our detachable rims referred to will be of ⅛ iron 1¾ wide and drawn together at one point by two screws, one on either side of felloe. It will be studded with calks in two rows." [34]

Figure 29. — Mr. and Mrs. Frank Duryea examining vehicle in the Smithsonian Institution before restoration.

January 18, 1894, was a day of triumph for Frank Duryea. Writing Charles about his success the next day he said, "Took out carriage again last night and gave it another test about 9 o'clock." The only difficulty he mentioned was a slight irregularity in the engine, caused by the tiny leather pad in the exhaust-valve mechanism falling out. [35] Speaking of this trip, Frank recalled in 1956:

When I got this car ready to run one night, I took it out and I had a young fellow with me; I thought I might need him to help push in case the car didn't work.... We ran from the area of the shop where it was built down on Taylor Street. We started out and ran up Worthington Street hill, [36] on top of what you might call "the Bluff" in Springfield. Then we drove along over level roads from there to the home of Mr. Markham who lived with his son-in-law, Will Bemis, and there we refilled this tank with water. [At this point he was asked if it was pretty well emptied by then.] Yes, I said in my account of it that when we got up there the water was boiling furiously. Well, no doubt it was. We refilled it and then we turned it back and drove down along the Central Street hill and along Maple, crossed into State Street, dropped down to Dwight, went west along Dwight to the vicinity where we had a shed that we could put the car in for the night. During that trip we had run, I think, just about six miles, maybe a little bit more. That was the first trip with this vehicle. It was the first trip of anything more than a few hundred yards that the car had ever made.

[Pg 27]

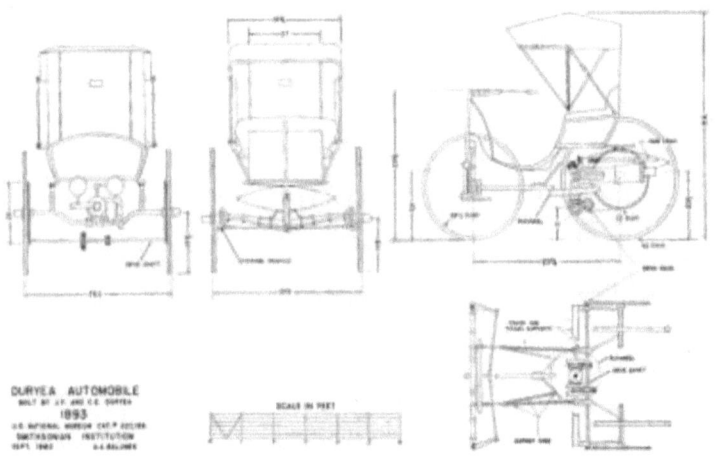

[Pg 28] Now Frank could give demonstration rides with the motor carriage, hoping to encourage more investors to back future work. Cautious Mr. Markham finally got his ride, though Frank had to assure him that the engine of the brakeless vehicle would hold them back on any hill they would descend. The carriage on which he had spent so many hours was to see little use after that. Its total mileage is probably less than a hundred miles. Little additional work is known to have been performed on the carriage after January 1894; there is, however, a letter [37] Frank sent his brother on January 19 which tells of contemplated muffler improvements. Another message was dispatched to Charles on March 22, mentioning the good performance of the phaeton on Harrison Avenue hill. [38] This was possibly the last run of the machine, for no further references have been discovered.

Frank spent the months of February and March in preparing drawings, some of which accompanied their first patent application, [39] while others were to be used in the construction of an improved, 2-cylinder carriage. Work on the new machine started in April. The old phaeton, in the absence of used-car lots, was put into storage in the Bemis barn. [40] Later, on the formation of the Duryea Motor Wagon Company in 1895, it was removed to the barn of D. A. Reed, treasurer of the company. [41] There it remained until

55

1920, when it was obtained by Inglis M. Uppercu and presented to the U.S. National Museum.

U.S. GOVERNMENT PRINTING OFFICE: 1967

For sale by the Superintendent of Documents, U.S. Government Printing Office
Washington, D.C. 20402 — Price 30 cents

Footnotes:

[1] S. H. Oliver, *Automobiles and Motorcycles in the U.S. National Museum* (U.S. National Museum Bulletin 213, Washington: Smithsonian Institution, 1957), p. 24.

[2] G. R. Doyle, *The World's Automobiles* (London: Temple Press Limited, 1959), p. 67.

[3] Recorded interview with Frank Duryea in the U.S. National Museum, November 9, 1956.

[4] Charles Duryea's statement to *Springfield Daily Republican*, April 14, 1937.

[5] Frank Duryea, *America's First Automobile* (Springfield, Mass.: Donald Macaulay, 1942), p. 4.

[6] Letter from Charles Duryea to Alfred Reeves, March 25, 1920; copy in Museum files.

[7] History notes dictated by Charles E. Duryea in the office of David Beecroft, editor of *Automobile Trade Journal*, on January 10, 1925. Copy in Museum files. Hereinafter, these notes are referred to as "history."

[8] Frank Duryea in statement made to the Senate Committee on Public Administration of Massachusetts, February 9, 1952.

[9] Duryea, op. cit. (footnote 5), p. 6.

[10] Copy of contract in Museum files.

[11] Affidavit of William Rattman, March 19, 1943, states that the Russell ledgers give that date.

[12] Recorded interview with Frank Duryea in U.S. National Museum, November 6, 1957.

[13] Letter from Frank Duryea to David Beecroft, November 15, 1924; copy in Museum files.

[14] Letter from Charles Duryea to C. W. Mitman, March 21, 1922; copy in Museum files.

[15] See "history" (footnote 7), p. 6.

[16] Duryea, op. cit. (footnote 5), p. 8.

[17] Copy of letter in Museum files.

[18] Duryea, op. cit. (footnote 5), p. 12.

[19] Letter from Frank Duryea to Charles Duryea, November 3, 1893, states that the engine could be run at 700 as well as 500 rpm. Copy in Museum files.

[20] Duryea, op. cit. (footnote 5), p. 14. Also in letter from Charles Duryea to C. W. Mitman, January 11, 1922; copy in Museum files.

[21] Letter from Charles Duryea to C. W. Mitman, January 11, 1922; also letter from Frank Duryea to David Beecroft, November 15, 1924. Copies in Museum files.

[22] Letter from Charles Duryea to F. A. Taylor, December 5, 1936, says he "thought" they had five teeth. Copy in Museum files.

[23] Frank later wrote his brother, January 1894, that he fixed the tank so it would not draw sediment from the bottom. Copy of letter in Museum files.

[24] The number of mufflers Frank Duryea constructed is not known. He wrote Charles, December 22, 1893, that he "will try a new muffler also."

[25] Selden Patent Evidence, vol. 9, p. 110.

[26] See "history" (footnote 7), p. 2. Charles wrote, "Some parts of these [referring to the batteries], like the jars, I had on hand for six or eight years, and did not need to buy."

[27] Ibid., p. 15.

[28] Ibid., p. 15

[29] Frank stated in this letter that the friction drum originally had two belts, forward and reverse, but since they tended to foul each other, he removed the reverse belt and left the other to serve for both directions. How the shipper fork might have handled two belts is not understood.

[30] As actually constructed there are only two gears on the secondary shaft. He obviously discovered that one gear secured to two clutches would serve for both forward and reverse. Space was also limited.

[31] Recorded interview with Frank Duryea in U.S. National Museum, November 9, 1956.

[32] Letter from Frank Duryea to Charles Duryea, November 8, 1893. Copy in Museum files.

[33] Frank Duryea, in a recorded interview in the U.S. National Museum on November 6, 1957, said that he believed these had been purchased from Rochester Rawhide Company.

[34] Letter from Frank Duryea to Charles Duryea, December 22, 1893. Also letter from Frank Duryea to David Beecroft, November 15, 1924. Copies in Museum files.

[35] Telling of the first use of the car in later days, Frank Duryea mentions the many noises and vibrations that accompanied the trip: the vibrating tiller, the tinny sounding muffler, the clattering chains. He later reported speeds of 3 mph in low gear and 8 mph in high gear.

[36] Letter from Frank Duryea to Charles Duryea, Jan. 19, 1894, says they went up hill via Summer and Armor Streets, then out Walnut to Bemis' at Central Street School.

[37] The letter read: "I have designed a new muffler and we will proceed to make it before long, in a day or two. Instead of one shell

⅛-inch thick I shall put a shell 1 1/16-inch thick inside another of equal thickness, but about 1 inch greater diameter i.e., one chamber within another so as to cause sound to turn corners to get out. Still another shell will be added if it prove insufficient, making it turn about again—taking care in each case to give ample room for expansion—outer one need not be more than 1/32 inch possibly. Will let two threaded rods with nuts hold heads on both or on three cases, if the 3d be essential."

[38] This letter gives further proof that the car never had a brake. Frank said the car came back down the hill with no brake, but that the engine held the vehicle back.

[39] Duryea, op. cit. (footnote 5), p. 37.

[40] It is possible that a few parts were removed at this time to be used on the two-cylinder car. The muffler may have been one of these, and even more likely, the governor parts. Charles Duryea wrote to C. W. Mitman December 27, 1921, stating that his younger brother Otho and a Henry Wells had put in a battery and gasoline in 1897 and started the engine. Because the chains were not on the car they could not attempt to operate it; but the engine ran too fast, and finally something broke, probably the engine frame, found to be broken during the recent restoration. Charles thought the engine ran too fast because some of the governor parts were already missing.

[41] Recorded interview with Frank Duryea in the U.S. National Museum, November 9, 1956. On the formation of the Duryea Motor Wagon Company, Mr. Markham was rewarded for his part of the venture. He had invested nearly $3000 in the work, and sold out his rights in the company for approximately a $2000 profit.

Text of Letter (page 14)

Dear Mr Mitman
24 Nov 1920

On the train I had some time to puzzle over that car. Been working nights to make up time lost in the day so did not hav much.

I made a sketch for you but did not show the spring that holds the circuit breaker in contact with the spark point. That thin finger was part of it. A spring was wound spirally--not helically--around the projecting end of the breaker pivot and the end of the spring hookt over the thin finger. See sketch herwith.

Just how the central end of the spring wire was fastened to the square of the pivot I do not kno. We did in some cases bore a hole thru and simply stick the spring thru but this put most of the action right at the bend in the wire and it broke quickly. So in other cases we fitted a light grooved spool or pulley and wound the spring around this and so avoided a sharp bend. If this was used it has been lost with the spring. A couple generations of boys playing in that barn was too many.

The Haynes steering sketch also worries me. If that vertical post came up thru that slot in the floor the crank had to be long as the sketch shows in order to get over to the driver conveniently. Then if he tried to make a complete circle with it he could not reach far enuf forward to do it easily. And he had to make a turn or two be cause H shows bevel gears of about same size so the post had to make same number of turns the worm made. Sketch herewith to illustrate my thought.

Text of Article (page 20)

NO USE FOR HORSES.

Springfield Mechanics Devise a New Mode of Travel.

Ingenious Wagon Now Being Made in This City for Which the Makers Claim Great Things.

A new motor carriage, which, if the preliminary tests prove successful as is expected, will revolutionize the mode of travel on highways, and do away with the horse as a means of transportation, is being made in this city. It is quite probable that within a short time one may be able to see an ordinary carriage in almost every respect, running along the streets or climbing country hills without visible means of propulsion. The carriage is being built by J. F.

Duryea, the designer and B. F. Markham, who have been at work on it for over a year. The vehicle was designed by C. E. Duryea, a bicycle manufacturer of Peoria, Ill., and he communicated his scheme to his brother, who is a practical machanic in this city.

The propelling power is furnished by a two-horse power gasoline motor situated near the rear axle and which, when started, runs continuously to the end of the trip, notwithstanding the number of times the carriage may be stopped. The speed of the motor is uniform, being about 500 revolutions a minute, and is so arranged that it gives a multiplied power for climbing hills and the lower the rate of speed the greater power is furnished by the motor. The slowest that the carriage can be driven is three miles an hour and the speed can be increased to fourteen or fifteen miles an hour. The power is transferred from the driving wheel of the motor, which runs horizontally with the main shaft by an endless friction belt running on a drum wheel. The belt is controlled by a lever within easy reach of the driver and is shifted along the drum wheel to increase or decrease the speed. The driving wheel is about twenty inches in diameter, having in its center a depression to which the belt is shifted to stop the carriage.

The carriage can be reversed by shifting the belt from the end of the drum, which gives the forward motion to the opposite side beyond the depression in the driving wheel. The power which has been transferred to the driving shaft from the motor is in turn transferred to the two rear wheels of the carriage by a combination gear and sprockets. An endless chain connects the sprockets on the carriage wheels to the sprocket wheels on the driving shaft. All of the motive power is located under the body of an ordinary phaeton, the hight of which is not increased by the machinery. The motor is started by a crank which is easily applied to a shaft in the rear of the carriage and the gasoline is ignited in the cylinder by electricity. An automatic device stops the flow of gasoline into the cylinder when the motor ceases running. The gasoline is carried in tanks, which hold about two gallons, and which will run the carriage for about eight hours. The wagon is guided by a bicycle bar, and the speed is also controlled by this bar.

The method employed in this is as follows: To start the carriage press the lever down; to reverse it throw the lever up and to guide the wagon turn the lever either to the right or left. The front axle instead of turning horizontally plays up and down, in order that the machinery may be on a level with the rear wheels, while the front wheels are set on the axle by a pivotal joint and are connected with the guiding lever by bars with ball bearings. The carriage complete weighs about 220 pounds, and the essential features are already covered by patents while others are pending.

It is estimated that the carriages can be sold for about $400, and a stock company will probably be formed to manufacture them.

Text of Patent Letter (page 24)

United States Patent Office.

CHARLES E. DURYEA, OF PEORIA, ILLINOIS.

ROAD-VEHICLE.

SPECIFICATION forming part of Letters Patent No. 540,648, dated June 11, 1895.

Application filed April 30, 1894. Serial No. 509,466. (No model.)

To all whom it may concern:

Be it known that I, Charles E. Duryea, a citizen of the United States, residing at Peoria, in the county of Peoria and State of Illinois, have invented new and useful Improvements in Road-Vehicles, of which the following is a specification.

The object of this invention is to produce a road vehicle which shall be self-propelled, not unduly heavy, simple and easy of control and comparatively inexpensive, together with such minor objects as will become hereinafter apparent.

The invention more particularly relates to the construction and arrangement of parts for constituting the driving gearing and to the means for controlling the action thereof; to an improved manner of mounting the front, or steering, wheels upon the front axle, and of mounting the said axle relative to the running gear frame, and to the means for effecting the steering; to the appliances for the support of the motor and driving mechanism in an advantageous and efficient manner, and, generally, to improved and simplified details of construction throughout the vehicle, all as will hereinafter be rendered more apparent, and the invention consists in constructions and combinations of parts, all substantially as will hereinafter fully appear and be set forth in the claims.

Reference is to be had to the accompanying drawings, in which—

Figure 1 is a sectional elevation from front to rear of the improved road-vehicle. Fig. 2 is a plan view of the running and driving gear, the vehicle-body being understood as removed. Fig. 3 is a front elevation of the vehicle. Fig. 4 is a perspective view of the support and suspension devices for the driving mechanism. Fig. 5 is a vertical sectional view, longitudinally, through the shiftable driving-gear, the controlling devices employed in conjunction with this mechanism being seen in side elevation. Figs. 6 and 7 show the above-mentioned controlling devices as in operative relations differing the one from the other and also from that of Fig. 5.

Similar letters of reference indicate corresponding parts in all of the views.

The parts will now be described in detail with reference to said drawings, and A represents the body which is spring supported on the frame, B, of the running gear. This frame, as shown, is rectangular, and has the body-supporting springs, B^2, similar to those found in common carriages. This frame has, affixed thereto, at its rear ends, sleeves, a, a, which loosely embrace the rear wheel axle, D, which is the driven axle of the vehicle. The axle, E, for the front wheels is centrally secured to the running gear frame, B, by the horizontal king-bolt, b, whereby such axle may have a swinging movement relative to the frame in a vertical plane, but it has no swinging movement horizontally, the wheels being swivel-mounted on the ends of this axle peculiarly, as will shortly hereinafter be set forth.

The body, as shown, is in the form of an inverted box, the motor, H, and driving gear being accommodated within the downwardly opening inclosure constituted thereby, and the body also has the upwardly open box-like forward extension, or pit, A^2, for the accommodation of the feet of the rider, the rider's seat being constituted by the top forward portion of the box body. Some other suitable design of body may, of course, be used in lieu of this one shown.

The front wheels, *d, d*, are hung to the front axle, E, so that the center of each wheel base is in a line coincident with the axis of the pivotal connection which is provided between the journals for the wheels and the axle, which arrangement practically destroys any tendency to deflection from the course that might otherwise arise from striking an obstacle, and so renders the steering easier. In order to effect this the axle is formed with yoked ends, the yoke members, *f, f*, being above and below the longitudinal line of the axle. The short journal, *g*, shown for each wheel, has at its inner end an upwardly and downwardly extended arm, *h*, which is return-bent to be loosely embraced by the axle yoke, *f, f*. The cone pointed screws, *c*, passed through the yoke members, *f*, and into sockets therefor in the arms, *h*, of the journals, *g*, constitute the means for the swivel connection between said parts. The lock-nuts, c^2, manifestly, are employed with utility in this connection.

It will be perceived that inasmuch as in the arrangement shown, the pivotal connections

www.ingramcontent.com/pod-product-compliance
Lightning Source LLC
Chambersburg PA
CBHW030504220526
45464CB00006B/2646